ANIMALS IN THEIR HABITATS

DESERT ANIMAL HABITATS

Amy Culliford

A Crabtree Roots Plus Book

Crabtree Publishing
crabtreebooks.com

School-to-Home Support for Caregivers and Teachers

This book helps children grow by letting them practice reading. Here are a few guiding questions to help the reader with building his or her comprehension skills. Possible answers appear here in red.

Before Reading:

- What do I think this book is about?
  - *I think this book is about animals.*
  - *I think this book is about animals that live in deserts.*
- What do I want to learn about this topic?
  - *I want to learn what animals live in deserts.*
  - *I want to know if birds live in deserts.*

During Reading:

- I wonder why...
  - *I wonder why sand is in most deserts.*
  - *I wonder why water is hard to find.*
- What have I learned so far?
  - *I have learned that some deserts have trees.*
  - *I have learned camels live in some deserts.*

After Reading:

- What details did I learn about this topic?
  - *I learned that some desert habitats have rivers.*
  - *I learned that camels can walk a long way and need very little water.*
- Read the book again and look for the vocabulary words.
  - *I see the word **sand** on page 6 and the word **reptiles** on page 14. The other vocabulary words are found on page 23.*

There are many kinds of animals that live in **desert habitats**.

Most animals that live in desert habitats live on land.

Sand can be found in most desert habitats. For most of the year, water is hard to find.

Some plants and trees can grow in desert habitats.

Some desert habitats have one or two large rivers.

Camels live in desert habitats. They can walk for many miles and need very little water.

Desert habitats are also home to many kinds of **reptiles**.

Some **crocodiles** live in desert habitats. They make their homes near rivers.

Vultures also live in desert habitats. They fly high in the sky to look for food.

Desert habitats are home to many animals!

Word List

Sight Words

also
and
animals
are
be
can
find
fly
food
for
found
grow
hard
have
high
home
homes
in
is
kinds
land
large
little
live
look
make
many
miles
most
near
need
of
on
one
or
plants
rivers
sky
some
that
the
their
there
they
to
trees
two
very
walk
water
year

Words to Know

camel

crocodile

desert habitats

reptile

sand

vulture

ANIMALS IN THEIR HABITATS

Amy Culliford

DESERT ANIMAL HABITATS

Written by: Amy Culliford
Designed by: Rhea Magaro
Series Development: James Earley
Proofreader: Kathy Middleton
Educational Consultant: Marie Lemke M.Ed.

Photographs: All images from Shutterstock

Crabtree Publishing

crabtreebooks.com 800.387.7650

 In Canada: We acknowledge the financial support of the Government of Canada through the Canada Book Fund for our publishing activities.

Printed in the USA/062024/CG20240201

Published in Canada
Crabtree Publishing
616 Welland Ave
St. Catharines, Ontario
L2M 5V6

Published in the United States
Crabtree Publishing
347 Fifth Ave
Suite 1402-145
New York, NY 10016

Library and Archives Canada Cataloguing in Publication
Available at Library and Archives Canada

Library of Congress Cataloging-in-Publication Data
Available at the Library of Congress

Hardcover: 978-1-0398-4462-9
Paperback: 978-1-0398-4543-5
Ebook (pdf): 978-1-0398-4617-3
Epub: 978-1-0398-4687-6
Read-Along: 978-1-0398-4757-6
Audio: 978-1-0398-4827-6